Hans-Friedrich Pfeiffer

Introduction to Einstein's Summation Notation

Introduction to Einstein's Summation Notation

Hans-Friedrich Pfeiffer

Impressum

Bibliografische Information der Deutschen Nationalbibliothek:
Die Deutsche Nationalbibliothek verzeichnet diese Publikation in der Deutschen
Nationalbibliografie; detaillierte bibliografische Daten sind im Internet über
http://dnb.dnb.de abrufbar.

© 2023 Hans-Friedrich Pfeiffer

Herstellung und Verlag: BoD – Books on Demand, Norderstedt

ISBN: 978-3-7412-9257-6

Contents

INTRODUCTION

The Einstein summation, also known as Einstein Summation Notation (ESN), is an effective tool for clearly presenting complicated mathematical expressions (sums of sums of sums). The ESN works with indices, i.e. "small" letters that are placed below or above an expression, for example a_{ij} or b^{ij}. The correct handling of these indices is essential for Einstein's summation, so in this book we will explain the basic properties and show the correct handling with numerous examples.

This publication is aimed in particular at physics students who want to deal with Albert Einstein's Special and General Theory of Relativity. In these areas in particular, correct handling of the ESN is essential.

Calculating with indices is also known as index acrobatics. Errors may have crept into this script as well. Should one be discovered, please send a message to **mathe@hans-friedrich-pfeiffer.de** . I will make errata available on

http://www.hans-friedrich-pfeiffer.de/Mathematik/en/ESN-Errata.pdf.

It is helpful if the reader is familiar with vector calculus and matrices. Knowledge of differential operators (gradient, curl, divergence) is also desirable. Knowledge of the areas mentioned is not absolutely necessary for the actual understanding, but we will use examples to explain how to work with the ESN and we get our examples from linear algebra and functional analysis.

CONVENTIONS

In the current literature, a distinction is made between covariant and contravariant vectors. This externally refers to contravariant vectors receiving upper indices, covariant lower indices. For our purposes this distinction is not necessary, so we will note all indices as sub indices.

In the following, $\left(\vec{e}_1, \vec{e}_2, \ldots, \vec{e}_n\right)$ denotes the orthonormal standard basis for the vector space $\mathbb{R}^n$. The components of the vector $\vec{e}_j$ are all 0 except for the j-th position, which contains a 1.

These vectors are perpendicular to each other (orthogonal) and all have the length 1 (orthonormalized). This basis system is also called **Cartesian**.

For matrices, we will number rows and columns starting with 0 - this happens because the so-called Minkoswki space is used in SRT/ART. This vector space is slightly different from the $\mathbb{R}^4$, the 2nd, 3rd and 4th components behave in the same way as the vector space $\mathbb{R}^3$.

In the Special and General Theory of Relativity, 4-dimensional spaces are considered. In our presentation, however, we will mostly allow any dimensions.

We will abbreviate Einstein's summation notation by **ESN**.

The sign "≡" stands for "corresponds to". We will use this character where we have an ESN on one side and show how this is represented as a summation.

TIP

Numerous equations are derived throughout the book. Read the equations in *both* directions, in particular from right to left. It is essential that relationships are recognized in a way that allows the reader to simplify the expression by paraphrasing.

EINSTEIN SUMMATION CONVENTION

In quite a lot of places you will encounter sums of sums of sums, expressions like

$$v_{;j'}^{i'} = \sum_{i=0}^{3}\sum_{j=0}^{3} \frac{\partial i'}{\partial i}\frac{\partial j}{\partial i'} \left(\sum_{l'}\sum_{k'} \left(\frac{\partial l'}{\partial j}\frac{\partial^2 i}{\partial l'\,\partial k'} v^{k'} + \frac{\partial i'}{\partial k'}\frac{\partial l'}{\partial j}\, v_{,l'}^{k'} \right) \right)$$

Sums that are lined up in this way quickly lead to confusion, so that essential points of such a summation are lost in the notation. We will see that the ESN can be used to formulate entire systems of equations in a single line.

Before we get into the actual ruleset, let's look at a few examples. The left side of the equal sign will mean Einstein's summation, the right side shows what exactly is meant by the left side. **In doing so, we must state in advance that only double-occurring indices are summed.**

The Einstein summation is characterized by the fact that you can leave out the sum-sign $\sum$, provided that some rules are follwoed.

In the examples below, a_i and u_i are real numbers and i is an index with i = 1, ..., n

(B 1) Example 1

The Einstein sum $\mathbf{a_i u_i}$ is defined as follows:

$$a_i\, u_i \equiv \sum_{i=1}^{n} a_i\, u_i = a_1\, u_1 + a_2\, u_2 + \ldots + a_n\, u_n$$

The double *(and ONLY double)* index "i" appearing on the left side of the equation is called a **bound** or **silent** index because it disappears after the summation is resolved.

$a_i\, u_i$ is just an abbreviated notation for the summation. So if it is clear that only double-occurring indices are summed up, the summation sign can be omitted.

(B 2) Example 2

$$a_{ij}u_i = \sum_{i=1}^{n} a_{ij}u_i = a_{1j}u_1 + a_{2j}u_2 + \ldots + a_{nj}u_n$$

Again, in this example, "i" is the bound index to sum over. The index "j" appears only once, it is called a **free** index. "j" can be freely chosen, but then remains fixed after its choice. So if "j" can take values from 1 to m, then the above expression results in a system of m summations:

$$a_{1j}u_1$$

$$a_{2j}u_2$$

$$\ldots$$

$$a_{mj}u_m$$

This shows that the ESN not only dispenses with the summation sign, but that a whole series of equations can be represented in one ESN.

In the last example a_{ij} could represent rows and columns of a matrix (with i as a row and j as a column), u_i could be the components of a vector. In this case the ESN represents the multplication of the matrix (a_{ij}) with the vector $(u_1, u_2, \ldots, u_n)$. It is important to understand that the Einstein summation **always refers only to components**. We will see later that many proofs in vector calculus get by with a pure representation as components, because if we carry out a proof for an (arbitrary) component, then the statement to be proved applies to all components. We shall also see that some of the proofs in the ESN are very simple and, for the main reason, concise.

We now come to some rules for the sum convention:

(R.1) Rule 1: only double-listed indices are totaled.

Example:

$$a_{ij}b_j = \sum_{j=1}^{n} a_{ij}b_j$$

In this example, the index "j" appears twice on the left. Therefore "j" is the summation index. The index "i", on the other hand, is free.

It is also possible for an expression to contain multiple bound indices:

$$a_{ij}b_{jk}c_k$$

In this example, "j" and "k" are bound indices - both(!) must be summed:

$$a_{ij}b_{jk}c_k = \sum_{j=1}^{n}\sum_{k=1}^{n} a_{ij}b_{jk}c_k$$

Here, too, there are m equations that the free index $"i"$ runs through.

(R.2) Rule 2: Only bound indices are allowed to be renamed:

Example:

$$a_{ij}b_j = a_{ik}b_k = a_{il}b_l$$

Here, in the second term, the double-occurring index $"j"$ was renamed to $"k"$, in the third term $"k"$ was then renamed to $"l"$.

On the other hand, the following ESN would be wrong:

$$a_{ij}b_j = a_{mj}b_j$$

Here the *free* index $"i"$ has been renamed to $"m"$.

Free indices may not be renamed.

If a bound index is to be renamed, then two points must be observed:

1. The new index name has not yet been used in the expression
2. All occurrences of the index to be renamed must be renamed.

(R.3) Rule 3: no index may appear more than twice *per term*.

Terms are added to each other. This means that all constructs that are multiplicatively connected count as one term.

Examples:

$a_{ij}b_j$ is a valid expression because *"j"* occurs twice and is therefore bound.

$a_{ij}b_j + c_{ij}d_j$ is also correct: we have two terms here and in *each* term "j" is the bound index.

$a_{ii}b_{ji}$ is not correct because *"i"* is used three times in the term.

(R.4) Rule 4: Any free index on the left-hand side of an Einstein summation equation must also appear as a free index on the right-hand side. Any free index on the right must also appear as a free index on the left. An index cannot be free on one side and a bound index on the other side.

An important note: the rule of consistency of free indices on both sides of an equation does NOT apply to bound indices. On the contrary: transformations, as we will discuss them later, serve to eliminate as many bound indices as possible. An example of this is considered below in (K.8).

Examples:

$x_i = a_{ij}b_j$ is correct because on both sides *"i"* is a free index. The index *"j"* is bound and occurs exactly twice in the term.

$x_i = a_{ki}b_{kj}x_j$ is also correct: on the right, "k" and "j" are bound indices, and the free index "i" on the left also appears as a free index on the right.

$x_i = a_{ki}b_{kj}c_m x_j$ is incorrect: the free index "m" appears on the right, which cannot be found on the left.

Identities and non-identities

Let's look at identities and non-identities using a few examples. In the case of non-identities in particular, we will investigate the reason why these cannot be valid.

(E.0) $x_i y_i = x_j y_j = x_k y_k$

Summation indices may be renamed arbitrarily according to (R.1). As trivial as this may seem, it is used extensively in the ESN.

For example, if two expressions need to be nested within each other - we'll see an example of this shortly - then renaming the bound index is necessary.

(E.1) $a_{ij}u_j = u_j a_{ij}$

Einstein sums are commutative, the order in a term, as well as the order of additively connected terms, is arbitrary. This is because in reality there is a summation behind it, and in this both the addition and the multiplication are commutative. However, we will discuss one restriction later: if operators (e.g. differential operators) are used in an Einstein summation, the operator cannot be interchanged with its argument.

In some cases, matrix multiplications are also described using Einstein sums: on the one hand, it should be noted that matrix multiplication is non-commutative and, on the other hand, that the result of an Einstein sum is not a matrix but represents individual components.

In the example above you can consider $A = \left(a_{ij}\right)$ as a matrix and $\boldsymbol{u}$ as a vector with the components $(u_1, u_2, \ldots, u_n)$. Then $a_{ij}u_j$ means row-by-row addition of a_{ij} with the components u_j : the result is the value of the ith component when matrix multiplying A by the vector $\boldsymbol{u}$. The result is a number, but not the vector!

We will cover matrices and Einstein summation in detail in a later chapter.

(E.2) $a_{ij}\left(x_i + y_j\right) \neq a_{ij}x_i + a_{ij}y_j$

On the left, "i" and "j" are the summation indices, so both must be summed.

On the right side of the expression $a_{ij}x_i$ is "i" the summation index. The index "j" is free. Exactly the opposite is the case in the expression $a_{ij}y_j$: here "j" is the summation index and "i" is free. Both sides as a whole are not consistent according to (R.4).

(E.3) $a_{ij}\left(x_j + y_j\right) = a_{ij}x_j + a_{ij}y_j$

Note here that x carries the same index as y. So in this case you can multiply out:

$$a_{ij}(x_j + y_j) \quad \equiv \quad \sum_{i=1}^{n}\sum_{j=1}^{n} a_{ij}(x_j + y_j)$$

$$= \quad \sum_{i=1}^{n}\sum_{j=1}^{n} (a_{ij}x_j + a_{ij}y_j)$$

$$= \quad \sum_{i=1}^{n}\sum_{j=1}^{n} a_{ij}x_j + \sum_{i=1}^{n}\sum_{j=1}^{n} a_{ij}y_j$$

$$\equiv \quad a_{ij}x_j + a_{ij}y_j$$

(E.4) $a_{ij}x_i y_j \neq a_{ij}x_j y_i$

The inequality should be obvious: you can't just swap the indices of objects.

(E.5) $(a_{ij} + a_{ji})x_i y_i \neq 2a_{ij}x_i y_i$

This formula is not valid: on the left hand side "j" is a summation index, but on the right hand side it is not - hence (R.4) is violated.

(E.6) On the other hand, the following applies:

$(a_{ij} + a_{ji})x_i x_j = 2a_{ij}x_i x_j$

We have:

$$
\begin{aligned}
\left(a_{ij} + a_{ji}\right)x_i x_j \quad &\equiv \quad \sum_{i=1}^{n}\sum_{j=1}^{n}\left(a_{ij} + a_{ji}\right)x_i x_j \\[4pt]
&= \; (a_{11} + a_{11})x_1 x_1 + (a_{12} + a_{21})x_1 x_2 + \cdots \\
&\qquad\qquad + (a_{1n} + a_{n1})x_1 x_n \\[4pt]
&+ \; (a_{21} + a_{12})x_2 x_1 + (a_{22} + a_{22})x_2 x_2 + \cdots \\
&\qquad\qquad + (a_{2n} + a_{n2})x_2 x_n \\[4pt]
&+ \; \ldots \\[4pt]
&+ \; (a_{n1} + a_{1n})x_n x_1 + (a_{n2} + a_{2n})x_n x_2 + \cdots \\
&\qquad\qquad + (a_{nn} + a_{nn})x_n x_n \\[4pt]
&= \; 2(a_{11})x_1 x_1 + 2(a_{12} + a_{21})x_1 x_2 + \cdots \\
&\qquad\qquad + 2(a_{nn})x_n x_n \\[4pt]
&\equiv \; 2a_{ij}x_i x_j
\end{aligned}
$$

Since $(a_{il} + a_{li})x_i x_l = (a_{li} + a_{il})x_l x_i$ we can combine the expressions framed above and get the desired result.

Another proof is much simlier: rename "i" to "j" and "j" to "i". We may do this as both indices are summation indices. That's it.

(E.7) Replacements of terms

We have an equation of the form $T = b_{ij}y_i x_j$ and we want to exchange the term y_i by $y_i = a_{ij}x_j$. If we simply did this as a "copy and paste", we would get the invalid term $= b_{ij}a_{ij}x_j x_j$, because the term indicates four occurrences of the index "j". In order to be able to carry out a correct replacement, it is necessary to proceed as follows:

In the term $y_i = a_{ij}x_j$ we must rename the bounded index "j" and we get a result like $y_i = a_{ir}x_r$ (cf. (E.0)).

With $T = b_{ij}y_ix_j$ and $y_i = a_{ir}x_r$ we get:

$$T = b_{ij}\, y_i\, x_j = b_{ij}\, (a_{ir}\, x_r)\, x_j = b_{ij}\, a_{ir}\, x_r\, x_j$$

(E.8) $a_{ij}x_ix_j = a_{ji}x_ix_j$

We have:

$$
\begin{aligned}
a_{ij}\, x_i\, x_j \;=\;\; & a_{1j}\, x_1\, x_j && [\,sum\ up \\
& +\ a_{2j}\, x_2\, x_j + \ldots + a_{nj}\, x_n\, x_j && over\ i\,] \\[4pt]
=\;\; & a_{11}\, x_1\, x_1 && [\,write\ out \\
& +\ a_{12}\, x_1\, x_2 + \ldots + a_{1n}\, x_1\, x_n && the\ sum\,] \\[4pt]
+\;\; & a_{21}\, x_2\, x_1 && \\
& +\ a_{22}\, x_2\, x_2 + \ldots + a_{2n}\, x_2\, x_n && \\[4pt]
+\;\; & \ldots && \\[4pt]
+\;\; & a_{n1}\, x_n\, x_1 && \\
& +\ a_{n2}\, x_n\, x_2 + \ldots + a_{nn}\, x_n\, x_n && \\[4pt]
=\;\; & a_{j1}\, x_j\, x_1 && [\,apply\ ESN \\
& +\ a_{j2}\, x_j\, x_2 + \ldots + a_{jn}\, x_j\, x_n && "vertically"\,] \\[4pt]
=\;\; & a_{ji}\, x_j\, x_i && \\[4pt]
=\;\; & a_{ji}\, x_i\, x_j && [\,(E.1)\,]
\end{aligned}
$$

As in (E.6) we obtain the result by just renaming the bounded index "i" to "j" and "j" to "i".

(E.9) $\left(a_{ij} - a_{ji}\right)x_i x_j = 0$

We have:

$$\begin{aligned}
\left(a_{ij} - a_{ji}\right)x_i x_j \; &= \; a_{ij}x_i x_j - a_{ji}x_i x_j \quad [\,cf.\ E.3\,] \\
&= \; a_{ij}x_i x_j - a_{ij}x_i x_j \quad [\,cf\ E.8\,] \\
&= \; 0
\end{aligned}$$

KRONECKER SYMBOL δ_{ij}

The Kronecker symbol δ_{ij} occurs very frequently in tensor calculation and can be defined in different ways:

1. as a function:

$$\delta_{ij} = \left\{ \begin{array}{l} 1 \ \ if \ \ i = j \\ 0 \ \ else \end{array} \right\}$$

2. as the unit matrix:

$$\delta_{ij} \ = \ \begin{pmatrix} 1 & 0 & 0 & \ldots & 0 \\ 0 & 1 & 0 & \ldots & 0 \\ 0 & 0 & 1 & \ldots & 0 \\ \ldots & \ldots & \ldots & \ldots & \ldots \\ 0 & 0 & 0 & 1 & 0 \\ 0 & 0 & 0 & 0 & 1 \end{pmatrix}$$

3. As the scalar product of the Cartesian unit vectors:

$$\delta_{ij} = \vec{e}_i \cdot \vec{e}_j$$

If "i" does not equal "j", the vectors are orthogonal and the dot product is 0. However, if "i" = "j", then the dot product is the square root of its length 1.

What is δ_{ij} used for? Imagine, you have a finite list of elements $(x_1, x_2, \ldots, x_n)$. Then $\delta_{ij} x_j$ "picks up" only the element x_j as we will see in short. Besides, there exists the so-called Dirac-Delta-Distribution $\delta(x)$ which does "the same thing" with an infinite set.

Identities:

(K.1) $\delta_{ij} = \delta_{ji}$

The indices of the Kronecker symbol may be swapped. This follows directly from the definition of the Kronecker-symbol.

(K.2) $\delta_{ij} x_i x_j = x_i x_i = x_j x_j$

"i" and "j" are each bound indices over which to sum. Now, all terms where "i" != "j" become 0 as δ_{ij} takes 0 in this case.

$$
\begin{aligned}
\delta_{ij} x_i x_j \;\equiv\; & \sum_{i=1}^{n}\sum_{j=1}^{n} \delta_{ij} x_i x_j \\
= \;& \delta_{11} x_1 x_1 + \delta_{12} x_1 x_2 + \cdots + \delta_{1n} x_1 x_n \\
+ \;& \delta_{21} x_1 x_1 + \delta_{22} x_1 x_2 + \cdots + \delta_{2n} x_1 x_n \\
+ \;& \ldots \\
+ \;& \delta_{n1} x_1 x_1 + \delta_{n2} x_1 x_2 + \cdots + \delta_{nn} x_n x_n \\
= \;& \delta_{11} x_1 x_1 + \delta_{22} x_2 x_2 + \cdots + \delta_{nn} x_n x_n \\
= \;& x_1 x_1 + x_2 x_2 + \cdots + x_n x_n \\
\equiv \;& x_i x_i = x_j x_j
\end{aligned}
$$

We reach the last equals sign by renaming the summation index.

(K.3) $\delta_{ij}\delta_{ik} = \delta_{ji}\delta_{ik} = \delta_{jk}$ **(contraction von δ)**

The summation runs over the bound index "i": δ_{ij} becomes 1 only when "i" reaches the value of "j", the same happens with δ_{ik} , so that in total there is only a non-zero term in the summation for i = j and i = k. With this we can eliminate one δ.

Note: contraction makes a bound index "disappear".

This simple rule is very commonly used, so it deserves special attention.

(K.4) $\delta_{kk} = \delta_{11} + \delta_{22}+...+\delta_{nn} = n$

Be careful with this formula: δ_{kk} does not equal to 1, as one might believe by definition. The index "k" appears twice: the summation must therefore be observed. Which value δ_{kk} takes depends on "what" δ_{kk} is connected to.

Below we will get to know the Levi-Civita tensor of the third order, whose indices i, j and k can assume the values 1, 2 or 3, respectively. In this case we get:

$$\varepsilon_{ijk}\delta_{mm=}\varepsilon_{ijk}3 = 3\varepsilon_{ijk}$$

(K.5) $\delta_{ij}\delta_{ji} = \delta_{ji}\delta_{ij} = n$

We have:

$$\begin{aligned}
\delta_{ij}\delta_{ji} &= \delta_{ii} \quad [\,cf.\ K.3\,] \\
&= n \quad [\,cf.\ K.4\,]
\end{aligned}$$

(K.6) $x_j \delta_{ji} = x_i$ **(exchange of indices)**

Again, note that the summation index vanishes.

We have to observe Einstein's summation convention here, it is summed over "j":

$$x_j \delta_{ji} \equiv \sum_{j=1}^{n} x_j \delta_{ji} = x_1 \delta_{1i} + x_2 \delta_{2i} + \ldots + x_i \delta_{ii} + \ldots + x_n \delta_{ni} \equiv x_i$$

Nearly all terms vanish, only the term remaining $x_i \delta_{ii}$ does not vanish when j reaches the value i.

The same argument applies more generally to any indexed object of the form $T_{ipq\ldots z}$:

$$T_{\underline{i}pq\ldots z} \delta_{ij} = T_{\underline{j}pq\ldots z}$$

The exchange of indices is one of the most important identities that is used in many places and therefore deserves the greatest attention.

(K.7) $\delta_{ij} \delta_{kj} \delta_{in} = \delta_{kn}$

Applying (K.3) twice gives this result.

(K.8) $x_{jmk} \delta_{nk} = x_{jmn}$

The summation takes place exclusively via the index "k", since this appears twice on the left. The indices "j", "m" and "n" are free. Only if "k" has the value "n" the summand does not disappear.

It should be noted that in the equation the summation index "k" on the left disappears. With summation indices, this is not only allowed, but also desired. This does not apply to free indices, they must reappear on both sides of the equation, as we have already explained in (R.4).

(K.9) $\delta_{ii} = \delta_{jj} = n^2$

The equation follows from (K.5).

K.(10) $\delta_{ij}\delta_{ik}\delta_{jk} = n$

We have:

$$\begin{aligned}
\delta_{ij}\delta_{ik}\delta_{jk} &= \delta_{jk}\delta_{jk} && [\,cf.\ (K.3)\,] \\
&= \delta_{kk} && [\,cf.\ (K.3)\,] \\
&= n && [\,cf.\ (K.4)\,]
\end{aligned}$$

LEVI-CIVITA SYMBOL ϵ_{ijk}

The Levi-Civita symbol ϵ_{ijk} plays an important role in the ESN. It doesn't show up that often in SRT / ART, but more so in electrodynamics, particularly when dealing with the various differential operators, which we'll discuss later.

The Levi-Civita-symbol is defined as follows with $i, j, k \in \{1,2,3\}$:

$$\textbf{(LC Def.):} \quad \epsilon_{ijk} = \epsilon_{jki} = \epsilon_{kij} = 1$$
$$\epsilon_{ikj} = \epsilon_{jik} = \epsilon_{kji} = -1$$
$$\epsilon_{ikk} = \epsilon_{iki} = \epsilon_{kii} = 0$$

The indices "i", "j" and "k" permute and take numbers from 1 to 3. If the permutation is even, then ε takes the value 1, if the permutation is odd, ε takes the value -1. If an index appears twice, the value is 0. We have defined ε for three dimensions - it also exists for arbitrary dimensions, but this is not needed in this book.

If a permutation occurs in the string "ijkijk" from left to right, the permutation is even and therefore takes the value 1. However, if the permutation is only from right to left, then it is an odd permutation and takes the value -1.

The calculation of the Levi-Civita symbol can also be done as follows:

$$\epsilon_{jjk} = (j-i)(k-j)(i-k) \bmod 3$$

The Levi-Civita symbol can also be expressed as follows:

$$\epsilon_{ijk} = \det\begin{pmatrix} \delta_{i1} & \delta_{i2} & \delta_{i3} \\ \delta_{j1} & \delta_{j2} & \delta_{j3} \\ \delta_{k1} & \delta_{k2} & \delta_{k3} \end{pmatrix}$$

$$= \delta_{i1}\delta_{j2}\delta_{k3} + \delta_{i3}\delta_{j1}\delta_{k2} + \delta_{i2}\delta_{j3}\delta_{k1} - \delta_{i3}\delta_{j2}\delta_{k1} - \delta_{i1}\delta_{j3}\delta_{k2} - \delta_{i2}\delta_{j1}\delta_{k3}$$

This equation can be recalculated directly. Actually, all possible combinations of the indices (i,j,k) would have to be calculated (then there would be 3^3 = 27 equations), with the following consideration we can reduce this to 6 cases:

If i = j or j = k or k = i and thus leads to a double index, then there are (at least) two equal (row) vectors in the determinant. These are then linearly dependent and the determinant is therefore 0. This correspondends with the definition of ϵ_{ijk}, if an index occurs (at least) twice. This leaves only 6 cases in which i,j and k are different from each other. In the sum above, all terms are sorted by i, j, and k. For all terms with a (+) sign, it is noticeable that the order of the second index each have an even permutation, corresponding to an odd permutation for the terms with a (-) sign. Depending on which permutation the first index goes through: only one term remains, all others become 0. The remaining term then also points to the correct sign of the permutation type.

In particular, this identity can be used to prove some other identities.

The following equations are helpful:

(LC.1) $\quad \epsilon_{ijk} \;=\; \epsilon_{jki} \;=\; \epsilon_{kij}$ (even permutation)

$\qquad\quad \epsilon_{kji} \;=\; \epsilon_{jik} \;=\; \epsilon_{ikj}$ (odd permutation)

$\qquad\quad \epsilon_{ijk} \;=\; -\epsilon_{kji}$ (permutation type changes)

We can swap the indices, but it must be noted that changing the type of permutation (from even to odd and vice versa) entails a change in sign.

(LC.2): $\epsilon_{jik}\epsilon_{lmn} = -\epsilon_{ijk}\epsilon_{lmn}$

Follows (LC.1)

(LC.3): contraction of ε with δ:

ε and δ often come together within one term. Note that in the following equations, the ESN must be applied since each *"i"* represents a summation index:

$$\delta_{im}\epsilon_{ijk} = \epsilon_{mjk}$$

$$\delta_{im}\epsilon_{kij} = \epsilon_{kmj}$$

$$\delta_{im}\epsilon_{jki} = \epsilon_{jkm}$$

These equations become clear, considering that you sum up over "i" and δ_{im} takes the value 1 only if i = m. Compare this with (K.6).

(LC.4): Double contraction of ε with δ:

$$\delta_{ij}\epsilon_{ijk} = \epsilon_{iik} = 0$$

Only for i=j δ_{ij} is not null 0, but exactly then ε has a double index and takes the value 0 by definition. Another way to see this identity is to apply (LC.3), equation 1 with m=j

(LC.5): $\epsilon_{ijk}\epsilon_{lmn} = \delta_{il}\delta_{jm}\delta_{kn} + \delta_{im}\delta_{jn}\delta_{kl} + \delta_{in}\delta_{jl}\delta_{km} - \delta_{il}\delta_{jn}\delta_{km} - \delta_{in}\delta_{jm}\delta_{kl} - \delta_{im}\delta_{jl}\delta_{kn}$

Note that the index (i j k) is passed through while the index (l m n) is permuted. This is one of the most important formulas for ϵ_{ijk} and is often used to simplify expressions.

To prove this identity, we exploit that the Levi-Civita symbol can be represented as a determinant (LC.0):

$$\epsilon_{ijk}\epsilon_{lmn} = \det\begin{pmatrix} \delta_{i1} & \delta_{i2} & \delta_{i3} \\ \delta_{j1} & \delta_{j2} & \delta_{j3} \\ \delta_{k1} & \delta_{k2} & \delta_{k3} \end{pmatrix} * \det\begin{pmatrix} \delta_{l1} & \delta_{l2} & \delta_{l3} \\ \delta_{m1} & \delta_{m2} & \delta_{m3} \\ \delta_{n1} & \delta_{n2} & \delta_{n3} \end{pmatrix}$$

According to the determinant theorem

$$det(A) * det(B) = det(A * B) \quad \text{we can write:}$$

$$\epsilon_{ijk}\epsilon_{lmn} = det\left[\begin{pmatrix} \delta_{i1} & \delta_{i2} & \delta_{i3} \\ \delta_{j1} & \delta_{j2} & \delta_{j3} \\ \delta_{k1} & \delta_{k2} & \delta_{k3} \end{pmatrix} * \begin{pmatrix} \delta_{l1} & \delta_{l2} & \delta_{l3} \\ \delta_{m1} & \delta_{m2} & \delta_{m3} \\ \delta_{n1} & \delta_{n2} & \delta_{n3} \end{pmatrix}\right]$$

Now we're going to take advantage of that $(A) = det(A^T)$. We transpose the second matrix and get:

$$\epsilon_{ijk}\epsilon_{lmn} = det\left[\begin{pmatrix} \delta_{i1} & \delta_{i2} & \delta_{i3} \\ \delta_{j1} & \delta_{j2} & \delta_{j3} \\ \delta_{k1} & \delta_{k2} & \delta_{k3} \end{pmatrix} * \begin{pmatrix} \delta_{l1} & \delta_{m1} & \delta_{n1} \\ \delta_{l2} & \delta_{m2} & \delta_{n2} \\ \delta_{l3} & \delta_{m3} & \delta_{n3} \end{pmatrix}\right]$$

We now perform matrix multiplication and get:

$$\epsilon_{ijk}\epsilon_{lmn}$$
$$= det\begin{pmatrix} \delta_{i1}\delta_{l1} + \delta_{i2}\delta_{l2} + \delta_{i3}\delta_{l3} & \delta_{i1}\delta_{m1} + \delta_{i2}\delta_{m2} + \delta_{i3}\delta_{m3} & \delta_{i1}\delta_{n1} + \delta_{i2}\delta_{n2} + \delta_{i3}\delta_{n3} \\ \delta_{j1}\delta_{l1} + \delta_{j2}\delta_{l2} + \delta_{j3}\delta_{l3} & \delta_{j1}\delta_{m1} + \delta_{j2}\delta_{m2} + \delta_{j3}\delta_{m3} & \delta_{j1}\delta_{n1} + \delta_{j2}\delta_{n2} + \delta_{j3}\delta_{n3} \\ \delta_{k1}\delta_{l1} + \delta_{k2}\delta_{l2} + \delta_{k3}\delta_{l3} & \delta_{k1}\delta_{m1} + \delta_{k2}\delta_{m2} + \delta_{k3}\delta_{m3} & \delta_{k1}\delta_{n1} + \delta_{n2}\delta_{m2} + \delta_{k3}\delta_{n3} \end{pmatrix}$$

By transposing the second matrix, we have managed to write down the summands resulting from the matrix multiplication in ESN. The summand $\delta_{i1}\delta_{l1} + \delta_{i2}\delta_{l2} + \delta_{i3}\delta_{l3}$ can be rewritten as $\delta_{is}\delta_{ls}$, where it should be noted that s must be summed. If we proceed analogously with all other summands, then we get:

$$\epsilon_{ijk}\epsilon_{lmn} = det\begin{pmatrix} \delta_{is}\delta_{ls} & \delta_{is}\delta_{ms} & \delta_{is}\delta_{ns} \\ \delta_{js}\delta_{ls} & \delta_{js}\delta_{ms} & \delta_{js}\delta_{ns} \\ \delta_{ks}\delta_{ls} & \delta_{ks}\delta_{ms} & \delta_{ks}\delta_{ns} \end{pmatrix}$$

We now use the contraction rule of the Kronecker symbol (K.3):
$$\delta_{is}\delta_{ls} = \delta_{il}$$

This applied to all matrix entries gives:

$$\epsilon_{ijk}\epsilon_{lmn} = det\begin{pmatrix} \delta_{il} & \delta_{im} & \delta_{in} \\ \delta_{jl} & \delta_{jm} & \delta_{jn} \\ \delta_{kl} & \delta_{km} & \delta_{kn} \end{pmatrix}$$

The direct calculation of the determinant then gives the claimed equation.

The equation is interesting in several respects: on the one hand, a connection is established between δ and ε, especially when two ε are multiplied. Some of the following relations between ε and δ refer to the representation of two ε-products by the determinant above.

Make yourself clear: the equation $\epsilon_{ijk}\,\epsilon_{lmn}$ represents $3^6 = 729$ equations! There is no more elegant way of putting this!

(LC.6): $\epsilon_{ijk}\epsilon_{lmk} = \delta_{il}\delta_{jm} - \delta_{im}\delta_{jl}$

Let's go into this evidence in a little more detail, as there's a stumbling block here that we'd like to examine in more detail.

First we use (LC.5) with n = k. We obtain:

$$\epsilon_{ijk}\epsilon_{lmk} = \delta_{il}\delta_{jm}\delta_{kk} + \delta_{im}\delta_{jk}\delta_{kl} + \delta_{ik}\delta_{jl}\delta_{km} - \delta_{il}\delta_{jk}\delta_{km} \\ - \delta_{ik}\delta_{jm}\delta_{kl} - \delta_{im}\delta_{jl}\delta_{kk}$$

Now comes the stopping stone: We want to replace δ_{kk}. We take up the discussion from (K.4) again and see that δ_{kk} takes the value 3, because the indices of δ_{kk} are running from 1 to 3.

If we would take $\delta_{kk} = 1$, so we would get a wrong result by one sign! The other way around: if an index acrobatic gives a result that is wrong by the sign, then you should check whether δ_{kk} appears in the expression!

So we use (K.4) that $\delta_{kk} = 3$ and we get:

$$\epsilon_{ijk}\epsilon_{lmk} = \delta_{il}\delta_{jm}3 + \delta_{im}\delta_{jk}\delta_{kl} + \delta_{ik}\delta_{jl}\delta_{km} - \delta_{il}\delta_{jk}\delta_{km} \\ - \delta_{ik}\delta_{jm}\delta_{kl} - \delta_{im}\delta_{jl}3$$

We use (K.3) and get:

$$\begin{aligned}
\epsilon_{ijk}\,\epsilon_{lmk} \;&=\; \delta_{il}\delta_{jm}3 + \delta_{im}\delta_{jl} + \delta_{im}\delta_{jl} - \delta_{il}\delta_{jm} \\
&\qquad - \delta_{il}\delta_{jm} - \delta_{im}\delta_{jl}3 \\[4pt]
&=\; 3\delta_{il}\delta_{jm} - \delta_{il}\delta_{jm} - \delta_{il}\delta_{jm} - 3\delta_{im}\delta_{jl} \qquad [\text{reorder}] \\
&\qquad - \delta_{im}\delta_{jl} - \delta_{im}\delta_{jl} \\[4pt]
&=\; \delta_{il}\delta_{jm} - \delta_{im}\delta_{jl}
\end{aligned}$$

(LC.7) $\epsilon_{ijk}\epsilon_{mnk} = \delta_{im}\delta_{jn} - \delta_{in}\delta_{jm}$

The proof of this is completely analogous to (LC.6)

(LC.8) $\epsilon_{ijk}\epsilon_{imn} = \delta_{jn}\delta_{km} - \delta_{kn}\delta_{jm}$

The proof of this is completely analogous to (LC.6)

(LC.9) The following equations apply:

$$\epsilon_{ijk}\epsilon_{imn} = \epsilon_{jki}\epsilon_{imn} = \epsilon_{kij}\epsilon_{imn} = \delta_{jn}\delta_{km} - \delta_{kn}\delta_{jm}$$

$$\epsilon_{ijk}\epsilon_{mni} = \epsilon_{jki}\epsilon_{mni} = \epsilon_{kij}\epsilon_{mni} = \delta_{jn}\delta_{km} - \delta_{kn}\delta_{jm}$$

$$\epsilon_{ijk}\epsilon_{nim} = \epsilon_{jki}\epsilon_{nim} = \epsilon_{kij}\epsilon_{nim} = \delta_{jn}\delta_{km} - \delta_{kn}\delta_{jm}$$

By exchanging the indices (observe LC.1!), numerous other equations can be set up. Now how do you know how to swap the indices correctly? We have already indicated this in the definition of ε: write the indices in a string, i.e. "ijkijk" and "imnimn" and check whether an exchange from right to left is contained. If so, you can exchange, if no, then you can exchange, but the sign must then be changed.

Tip: don't memorize all possible combinations. Reduce these to (LC.6) to (LC.8) and, for unknown index combinations, see whether you can get one of the known forms by permutation or renaming.

(LC.10) $\epsilon_{ijk}\epsilon_{ljk} = 2\delta_{li}$

We have :

$$
\begin{aligned}
\epsilon_{ijk}\epsilon_{ljk} &= \delta_{il}\delta_{jj} - \delta_{ij}\delta_{jl} \quad && [\ (LC.6)\ with\ m = j\ ] \\
&= 3\delta_{il} - \delta_{ij}\delta_{jl} \quad && [\ (K.4)\ with\ n = 3\ ] \\
&= 3\delta_{il} - \delta_{il} \quad && [\ (K.3)\ ] \\
&= 2\delta_{il}
\end{aligned}
$$

(LC.11) $\epsilon_{ijk}\epsilon_{ijk} = 6$

$$
\begin{aligned}
\epsilon_{ijk}\epsilon_{ijk} &= 2\delta_{ii} \quad && [\ (LC.10)\ with\ l = i\ ] \\
&= 6 \quad && [\ (K.3)\ ]
\end{aligned}
$$

(LC.12) $\epsilon_{imn}\epsilon_{jmn} = 2\delta_{ij}$ (**double contraction od ε**)

Immediately follows from (LC.10)

(LC.13) $\epsilon_{ijk}\epsilon_{jnl}\epsilon_{mil} = \epsilon_{nmk}$ **(triple contraction of ϵ)**

We have:

$$
\begin{aligned}
\epsilon_{ijk}\epsilon_{jnl}\epsilon_{mil}
&= \epsilon_{ijk}\left(\delta_{jm}\delta_{ni} - \delta_{ji}\delta_{mn}\right) && [\,(1)\,] \\
&= \epsilon_{ijk}\delta_{jm}\delta_{ni} - \epsilon_{ijk}\delta_{ji}\delta_{mn} && \\
&= \epsilon_{njk}\delta_{jm} - \epsilon_{jjk}\delta_{nm} && [\,(LC.3)\,] \\
&= \epsilon_{njk}\delta_{jm} && [\,\epsilon_{jjk} = 0\,] \\
&= \epsilon_{nmk} && [\,(LC.3)\,]
\end{aligned}
$$

   (1) *(LC.6) with i=j, j=n, k=l, l=m, m=i*

(LC.14) $\epsilon_{ijk}\epsilon_{klm} = \delta_{il}\delta_{jm} - \delta_{im}\delta_{jl}$

The proof could be carried out analogously to (LC.6), but we will also show another possibility of proof:

According to (LC.5) (with l = i):

$$
\begin{aligned}
\epsilon_{ijk}\epsilon_{imn}
&= \delta_{ii}\delta_{jm}\delta_{kn} + \delta_{im}\delta_{jn}\delta_{ki} + \delta_{in}\delta_{ji}\delta_{km} \\
&\quad - \delta_{ii}\delta_{jn}\delta_{km} - \delta_{in}\delta_{jm}\delta_{ki} \\
&\quad - \delta_{im}\delta_{ji}\delta_{kn} \\
&= 3\delta_{jm}\delta_{kn} + \delta_{jn}\delta_{km} + \delta_{jn}\delta_{km} - 3\delta_{jn}\delta_{km} \\
&\quad - \delta_{jm}\delta_{kn} - \delta_{jm}\delta_{kn} \\
&= \delta_{jm}\delta_{kn} - \delta_{jn}\delta_{km}
\end{aligned}
$$

Now swap the indices as follows:

l $\rightarrow$ k $\rightarrow$ j $\rightarrow$ i , m $\rightarrow$ l, n $\rightarrow$ m

This gives us:

$$\epsilon_{kij}\epsilon_{klm} = \epsilon_{ijk}\epsilon_{klm} = \delta_{il}\delta_{jm} - \delta_{im}\delta_{jl}$$

In the transformations we have used (K.1), (K.3) and (K.4) several times.

(LC.15) $\epsilon_{ijk}\epsilon_{lmk} = \delta_{il}\delta_{jm} - \delta_{im}\delta_{jl}$

With (LC.5) we get:

$$\begin{aligned}
\epsilon_{ijk}\epsilon_{lmk} &= \delta_{il}\delta_{jm}\delta_{kk} + \delta_{im}\delta_{jk}\delta_{kl} + \delta_{ik}\delta_{jl}\delta_{km} - \delta_{il}\delta_{jk}\delta_{km} \\
&\quad - \delta_{ik}\delta_{jm}\delta_{kl} - \delta_{im}\delta_{jl}\delta_{kk} \\
&= 3\delta_{il}\delta_{jm} + \delta_{im}\delta_{jl} + \delta_{im}\delta_{jl} - \delta_{il}\delta_{jm} - \delta_{il}\delta_{jm} \\
&\quad - 3\delta_{im}\delta_{jl} \\
&= \delta_{il}\delta_{jm} - \delta_{im}\delta_{jl}
\end{aligned}$$

(LC.16) $\epsilon_{ijk}\epsilon_{ljk} = 2\delta_{il}$

With (LC.5) we get:

$$\begin{aligned}
\epsilon_{ijk}\epsilon_{ljk} &= \delta_{il}\delta_{jj} - \delta_{ij}\delta_{jl} \\
&= 3\delta_{il} - \delta_{il} \\
&= 2\delta_{il}
\end{aligned}$$

(LC.17) $\epsilon_{ijk}\epsilon_{jnl}\epsilon_{ilm} = \epsilon_{nmk}$

We have:

$$\epsilon_{ijk}\epsilon_{jnl}\epsilon_{ilm} = \epsilon_{ijk}\epsilon_{jnl}\epsilon_{mil} \qquad [\,\epsilon_{ilm} = \epsilon_{mil}\,]$$

$$\begin{aligned}
&= \quad \epsilon_{ijk}\left(\delta_{jm}\delta_{ni} - \delta_{ji}\delta_{nm}\right) &&[\ (LC.16)\ ] \\
&= \quad \epsilon_{ijk}\delta_{jm}\delta_{ni} - \epsilon_{ijk}\delta_{ji}\delta_{nm} &&[\ multiplied\ out\ ] \\
&= \quad \delta_{jm}\epsilon_{njk} - \delta_{nm}\epsilon_{jjk} &&[\ (LC.2)\ ] \\
&= \quad \epsilon_{nmk}
\end{aligned}$$

VECTOR CALCULATION IN INDEX NOTATION

In this section we look at how vector calculus can be performed in index notation. It will turn out that some proofs can be formulated very simply using index notation.

Let be $\vec{a}$ and $\vec{b}$ vectors with the components $(a_1, a_2, \ldots, a_n)$ respectively $\left(\vec{b_1}, \vec{b_2}, \ldots, \vec{b_3}\right)$.

Representation of a vector

Is $\left(\vec{g_1}, \vec{g_2} \ldots, \vec{g_n}\right)$ an (arbitrary) basis, then each vector can be represented as a linear combination of the basis vectors:

$$\vec{a} = \sum_{i=1}^{n} a_i g_i$$

In index notation we get: $\vec{a} = a_i g_i$

Vectors are added component by component:

$$\vec{a} + \vec{b} = \begin{pmatrix} a_1 \\ a_2 \\ \ldots \\ a_n \end{pmatrix} + \begin{pmatrix} b_1 \\ b_2 \\ \ldots \\ b_n \end{pmatrix} = \begin{pmatrix} a_1 + b_1 \\ a_2 + b_2 \\ \ldots \\ a_n + b_n \end{pmatrix}$$

In index notation we get for the component "j":

$$\left(\vec{a} + \vec{b}\right)_j = a_j + b_j$$

Please note: in the index notation we can only show how the component "j" is calculated!

Scalar product in index notation

Using the previous notations, we define the scalar product to the basis $\left(\vec{g}_1, \vec{g}_2 \ldots, \vec{g}_n \right)$ as follows:

$$
\begin{aligned}
\vec{a} \cdot \vec{b} \;=\; & \sum_{i=1}^{n} a_i g_i \;*\; \sum_{j=1}^{n} b_j g_j \\[2mm]
=\; & (a_1 g_1 + a_2 g_2 + \ldots + a_n g_n)(b_1 g_1 \\
& \qquad\qquad\qquad + b_2 g_2 + \ldots + b_n g_n) \\[2mm]
=\; & a_1 g_1 b_1 g_1 \\
& + a_1 g_1 b_2 g_2 + \ldots + a_1 g_1 b_n g_n \\
& + a_2 g_2 b_1 g_1 \\
& + a_2 g_2 b_2 g_2 + \ldots + a_2 g_2 b_n g_n \\
& + \ldots \\[2mm]
& + a_n g_n b_1 g_1 \\
& + a_n g_n b_2 g_2 + \ldots + a_n g_n b_n g_n \\[2mm]
=\; & \sum_{i=1}^{n} \sum_{j=1}^{n} a_i g_i b_j g_j \\[2mm]
=\; & \sum_{i=1}^{n} \sum_{j=1}^{n} a_i b_j g_i g_j \\[2mm]
=\; & \sum_{i=1}^{n} \sum_{j=1}^{n} a_i b_j g_{ij} \qquad\qquad [\, g_{ij} := g_i g_j \,] \\[2mm]
=\; & a_i b_j g_{ij}
\end{aligned}
$$

In the penultimate step we defined the scalar products of the basis vectors $\vec{g}_i$ and $\vec{g}_j$ as g_{ij}. (g_{ij}) can be viewed as a matrix and is called the metric tensor. If it is a Cartesian basis with unit vectors, then (g_{ij}) is the unit matrix. This can be described as a Kronecker delta, so the scalar product for the Cartesian basis system takes the form:

$$\vec{a} \cdot \vec{b} = a_i b_j \delta_{ij} = a_i b_i$$

Cross product in index notation

In the three-dimensional real vector space, the cross product of two vectors is defined as follows:

$$\vec{a} \times \vec{b} := \begin{pmatrix} a_2 b_3 - a_3 b_2 \\ a_3 b_1 - a_1 b_3 \\ a_1 b_2 - a_2 b_1 \end{pmatrix}$$

The result is a vector that is perpendicular to the plane defined by the vectors $\vec{a}$ and $\vec{b}$. Its length corresponds to the area of $\vec{a}$ and $\vec{b}$.

Referred to the standard orthonormal basis $\left(\vec{e}_1, \vec{e}_2, \vec{e}_3\right)$ the vector can also be represented like this:

$$\vec{a} \times \vec{b} = (a_2 b_3 - a_3 b_2)\vec{e}_1 + (a_3 b_1 - a_1 b_3)\vec{e}_2 + (a_1 b_2 - a_2 b_1)\vec{e}_3$$

In index notation it looks like this:

$$\textbf{(V.1)} \quad \left(\vec{a} \times \vec{b}\right)_k = \varepsilon_{ijk} a_i b_j$$

with i,j and k from 1 to 3, where $\left(\vec{a} \times \vec{b}\right)_k$ means the k-th component. In this case, k (as the component to be considered) is to be chosen as a fixed value and only i and j are summed. When forming the sum, it should be noted that all summands where i=j or j=k or k=i vanish after the definition of ε.

The cross product is anti-commutative, it holds:

(V.2) $\vec{a} \times \vec{b} = -\vec{b} \times \vec{a}$:

We have:

$$\left(\vec{a} \times \vec{b}\right)_k = \varepsilon_{ijk} a_i b_j$$

$$\left(-\vec{b} \times \vec{a}\right)_k = -\varepsilon_{ijk} b_i a_j$$

In the last equation we swap i and j: j becomes i and i becomes j: we are allowed to do this since both indices are summation indices. So it follows:

$$\left(-\vec{b} \times \vec{a}\right)_k = -\varepsilon_{jik} a_j b_i$$

However, according to (LC.1, third equation) $\varepsilon_{jik} = -\varepsilon_{ijk}$, so that we get now

$$\left(-\vec{b} \times \vec{a}\right)_k = -\varepsilon_{jik} a_j b_i = \varepsilon_{ijk} a_j b_i = \left(\vec{a} \times \vec{b}\right)_k$$

Now, we want to show how to proove some weel-known identities using ESN. The main idea behind is that we proof an identity for an arbitrary component. If the identity holds for one componenet, it will hold for all components.

Graßmann identity

$$\vec{a} \times \left(\vec{b} \times \vec{c}\right) = \left(\vec{a} \cdot \vec{c}\right) \cdot \vec{b} - \left(\vec{a} \cdot \vec{b}\right) \cdot \vec{c}$$

The proof of the Graßmann identity in index notation shows this for an (arbitrary) component "k":

We have:

$$
\begin{aligned}
\left[\vec{a} \times \left(\vec{b} \times \vec{c}\right)\right]_k
&= \varepsilon_{ijk} a_i (b \times c)_j & &\text{[def.]} \\
&= \varepsilon_{ijk} a_i \varepsilon_{lmj} b_l c_m & &\text{[def.]} \\
&= \varepsilon_{ijk} \varepsilon_{lmj} a_i b_l c_m & &\text{[reorder]} \\
&= (\delta_{kl}\delta_{im} - \delta_{km}\delta_{il}) a_i b_l c_m & &\text{[(LC.6)]} \\
&= \delta_{kl}\delta_{im} a_i b_l c_m - \delta_{km}\delta_{il} a_i b_l c_m & & \\
&= a_i b_i c_k - a_i c_i b_k & &\text{[contract]} \\
&= \left(\vec{a} \cdot \vec{c}\right) b_k - \left(\vec{a} \cdot \vec{b}\right) c_k & & \\
&= [(a \cdot c)b - (a \cdot b)c]_k & &
\end{aligned}
$$

Lagrange identity

$$\left(\vec{a} \times \vec{b}\right)\left(\vec{c} \times \vec{d}\right) = \left(\vec{a} \cdot \vec{c}\right)\left(\vec{b} \cdot \vec{d}\right) - \left(\vec{a} \cdot \vec{d}\right)\left(\vec{b} \cdot \vec{c}\right)$$

For the k-th component we get:

$$
\begin{aligned}
\left[\left(\vec{a} \times \vec{b}\right)\left(\vec{c} \times \vec{d}\right)\right]_k
&= \left(\varepsilon_{ijk} a_i b_j\right)\left(\varepsilon_{lmk} c_l d_m\right) & (1)\\
&= \varepsilon_{ijk}\varepsilon_{lmk} a_i b_j c_l d_m & (2)\\
&= \left(\delta_{il}\delta_{jm} - \delta_{im}\delta_{jl}\right) a_i b_j c_l d_m & (3)\\
&= \delta_{il}\delta_{jm} a_i b_j c_l d_m - \delta_{im}\delta_{jl} a_i b_j c_l d_m & (4)\\
&= a_i b_j c_i d_j - a_m b_l c_l d_m & (5)\\
&= \left(\vec{a} \cdot \vec{c}\right)\left(\vec{b} \cdot \vec{d}\right) - \left(\vec{a} \cdot \vec{d}\right)\left(\vec{b} \cdot \vec{c}\right) & (6)
\end{aligned}
$$

(1) def. (2) reorder (3) LC.6 (4) multiply out

(5) contract (6) definition

MATRIX MULTIPLICATION IN INDEX NOTATION

For a matrix, we will count the rows and columns starting with 0. The reason is that the index notation is used in general and special relativity in particular. The associated 4-dimensional vector space (Minkowski space) consists of 4 vectors whose components are numbered with 0 for specific reasons.

Matrix multiplication can also be represented as Einstein summation. To do this, we first look at the matrix multiplication of two 3x3 matrices:

$$C = AB$$

$$\begin{pmatrix} a_{00} & a_{01} & a_{02} \\ a_{10} & a_{11} & a_{12} \\ a_{20} & a_{21} & a_{22} \end{pmatrix} \begin{pmatrix} b_{00} & b_{01} & b_{02} \\ b_{10} & b_{11} & b_{12} \\ b_{20} & b_{21} & b_{22} \end{pmatrix} =$$

$$\begin{pmatrix} a_{00}b_{00} + a_{01}b_{10} + a_{02}b_{20} & a_{00}b_{01} + a_{01}b_{11} + a_{02}b_{21} & a_{00}b_{02} + a_{01}b_{12} + a_{02}b_{22} \\ a_{10}b_{00} + a_{11}b_{10} + a_{12}b_{20} & a_{10}b_{01} + a_{11}b_{11} + a_{12}b_{21} & a_{10}b_{02} + a_{11}b_{12} + a_{12}b_{22} \\ a_{20}b_{00} + a_{21}b_{10} + a_{22}b_{20} & a_{20}b_{01} + a_{21}b_{11} + a_{22}b_{21} & a_{20}b_{02} + a_{21}b_{12} + a_{22}b_{22} \end{pmatrix} =$$

$$\begin{pmatrix} a_{0i}b_{i0} & a_{0i}b_{i1} & a_{0i}b_{i2} \\ a_{1i}b_{i0} & a_{1i}b_{i1} & a_{1i}b_{i2} \\ a_{2i}b_{i0} & a_{2i}b_{i1} & a_{2i}b_{i2} \end{pmatrix}$$

Here we have written the individual components as ESN.

Defining the matrix A by $A^{\mu}{}_{\nu}$, accordingly B by $B = B^{\mu}{}_{\nu}$ and $C = C^{\mu}{}_{\nu}$, then we get:

$$C^{\mu}{}_{\nu} = A^{\mu}{}_{\gamma} B^{\gamma}{}_{\nu} = B^{\gamma}{}_{\nu} A^{\mu}{}_{\gamma}$$

On the one hand, it should be noted that the multiplication in the ESN is commutative, since it only refers to <u>one component</u>. Second, that the summation index γ one of the multipliers is the <u>inner</u> index and the other is the <u>outer</u> index!

The result here is not a matrix but calculates for each row and column the value of the component $C^{\mu}{}_{\nu}$. For example, the component $C^2{}_1$ results as $A^2{}_{\gamma}\ B^{\gamma}{}_1 = B^{\gamma}{}_1\ A^2{}_{\gamma}$, summed up over γ. The Einstein notation always refers *only to components*, not to an entire object. If we want to denote a matrix with the Einstein notation, then we put this into brackets, for example $\left(C^{\mu}{}_{\nu}\right)$.

Now let's consider the case that $= A^T B$, i.e. the multiplication of the transposed matrix A by matrix B.

With

$$\left(A^{\mu}{}_{\nu}\right) = \begin{pmatrix} a_{00} & a_{01} & a_{02} \\ a_{10} & a_{11} & a_{12} \\ a_{20} & a_{21} & a_{22} \end{pmatrix}$$

We get:

$$\left(A^{\mu}{}_{\nu}\right)^T = \begin{pmatrix} a_{00} & a_{10} & a_{20} \\ a_{01} & a_{11} & a_{21} \\ a_{02} & a_{12} & a_{22} \end{pmatrix}$$

In component notation, the indices interchange as follows:

$$\left(A^{\mu}{}_{\nu}\right)^T = \left(A^{\nu}{}_{\mu}\right)$$

Let's see what comes out as a result of matrix multiplication $A^T B$:

$$C^{\mu}{}_{\nu} = \left(A^{\mu}{}_{\gamma}\right)^T B^{\gamma}{}_{\mu} = A^{\gamma}{}_{\mu} B^{\gamma}{}_{\mu}$$

Note that in this case you have to sum up over the *inner* index .

We now look at what happens at $C = A\,B^T$:

$$C^\mu{}_\nu = A^\mu{}_\gamma \left(B^\gamma{}_\nu\right)^T = A^\mu{}_\gamma B^\nu{}_\gamma$$

Here it is the *outer* index that is summed up!

Summary:

It is important that based on the location of the bound index - in our case this is γ – it can be recognized what kind of matrix multiplication is described by the ESN:

Matrix multiplication	in component notation	position of the summation index
AB	$A^\mu{}_\gamma B^\gamma{}_\nu$ bzw. $B^\gamma{}_\nu A^\mu{}_\gamma$	outer times inner index or inner times outer index
$A^T B$	$A^\gamma{}_\mu B^\gamma{}_\nu$ bzw. $B^\gamma{}_\nu A^\gamma{}_\mu$	inner times inner index
AB^T	$A^\mu{}_\gamma B^\nu{}_\gamma$ bzw. $B^\nu{}_\gamma A^\mu{}_\gamma$	outer times outer index

Inverse matrix:

A matrix A is invertible if there exists a matrix B with $AB = BA = E$, where E shall be the identity matrix. B is then the inverse matrix of

A , A is also the inverse matrix of B . The condition in component notation is then:

$$A^{\mu}{}_{\gamma} B^{\gamma}{}_{\nu} = \delta_{\mu\nu}$$

Let's look at a proof, which we do in component notation:

(M.1) $AB^{T} = A^{T} B^{T}$

Define $C = \left(C^{\mu}{}_{\nu} \right) = AB$

Then we get:

$$(AB)^{T} = \left(A^{\mu}{}_{\nu} \, B^{\nu}{}_{\mu} \right)^{T} = \left(C^{\mu}{}_{\nu} \right)^{T} = C^{\nu}{}_{\mu}$$

But $A^{T} = \left(A^{\nu}{}_{\mu} \right)$ and $B^{T} = \left(B^{\mu}{}_{\nu} \right)$, the indices have swapped.

It follows:

$$A^{T} B^{T} = A^{\nu}{}_{\mu} B^{\mu}{}_{\nu} = C^{\mu}{}_{\nu}$$

A comparison with $(AB)^{T} = C^{\mu}{}_{\nu}$ yields the result.

We want to pick up another point: how do you interpret an ESN equation? Let's take the following equation in ESN as an example:

$$a_{ij}x_j = b_i$$

A system of n equations is obviously set up here, which can be represented in matrix form as follows:

$$Ax = b,$$

where $= (a_{ij})$, $x = (x_1, x_2, \ldots x_n)$ and $b = (b_1, b_2, \ldots b_n)$ shall apply. This form should be familiar from linear algebra: it describes a linear system of equations.

But what about the following expression?

$$Q = a_{ij}x_i x_j$$

Here it is no longer clearly recognizable how this expression can be represented in matrix notation. In fact, the equation can, among other things, represent a so-called *quadratic form with a homogeneous multi-variable polynomial of the second order*. We explain this with an example. Consider:

$$10 = 3x^2 + y^2 - z^2 - 5xy - 6yz$$

The polynomial has several variables: x,y,z . *Homogeneous of the second order* now means that all terms (one also speaks of monomials here) must have degree 2:

For $3x^2$, y^2, z^2 this is obviously the case. For 5xy, the degrees of the individual variables are added together and this gives degree 2, the same for 6yz.

Such quadratic forms can always be represented as a matrix equation:

$$Q = x^T A x$$

Detailed (for n = 3):

$$Q = (x_1 \ x_2 \ x_3) \begin{pmatrix} a_{11} & a_{12} & a_{13} \\ a_{21} & a_{22} & a_{23} \\ a_{31} & a_{32} & a_{33} \end{pmatrix} \begin{pmatrix} x_1 \\ x_2 \\ x_3 \end{pmatrix}$$

For our example we get:

$$10 = (x_1 \ x_2 \ x_3) \begin{pmatrix} 3 & -5 & 0 \\ 0 & 1 & 6 \\ 0 & 0 & -2 \end{pmatrix} \begin{pmatrix} x_1 \\ x_2 \\ x_3 \end{pmatrix}$$

When translating an ESN back into a matrix equation, special attention must be paid to the fact that vectors may need to be transposed (so that they then become a row vector) so that the multiplication steps can be performed consistently. However, this cannot be seen directly in the ESN equation!

It is therefore important that you get a feeling for wether - and above all: how - an ESN with two indices can be represented as a matrix equation. As much as an ESN can help to represent entire collections of equations elegantly: when it comes to concrete calculations, matrix multiplication is easier to carry out. As a final remark one might add that a whole range of tensors in physics (these are multi-linear maps with a very specific transformation behaviour) can be and are represented in matrix form.

ESN AND PARTIAL DERIVATIVES

We now look at the ESN in partial derivatives.

For the beginning, here are some equations that have to do with partial derivatives. First, we note that the following equation holds:

$$\textbf{(PA.0)} \quad \frac{\partial x_p}{\partial x_k} = \delta_{pk}$$

With that in mind, let's now look at the following equations:

$$\textbf{(PA.1)} \quad \frac{\partial}{\partial x_k}\left(a_{ij}x_j\right) = a_{ij}\frac{\partial}{\partial x_k}x_j = a_{ij}\delta_{jk} = a_{ik}$$

We used (PA.0) and (K.6) here.

$$\textbf{(PA.2)} \quad \frac{\partial}{\partial x_k}\left(a_{11}x_1 + a_{12}x_2 + a_{13}x_3\right) = a_{1k} \qquad \text{for k = 1,2,3}$$

We have:

$$\frac{\partial}{\partial x_k}\left(a_{11}x_1 + a_{12}x_2 + a_{13}x_3\right)$$

$$= \quad a_{11}\frac{\partial}{\partial x_k}x_1 + a_{12}\frac{\partial}{\partial x_k}x_2 + a_{13}\frac{\partial}{\partial x_k}x_3$$

$$= \quad a_{11}\delta_{1k} + a_{12}\delta_{2k} + a_{13}\delta_{3k}$$

$$= \quad a_{1k}$$

Only one of the δ "survives".

(PA.3) $\quad \dfrac{\partial}{\partial x_k}\left(a_{ij}x_ix_j\right) = (a_{ik} + a_{ki})x_i$

We note the following:

1. $\sum_{i=1,j=1}^{n} mit\ i<>=k,j<>k\ a_{ij}\,x_ix_j$
   is independant of x_k , because k is explicitly excluded.

2. $\sum_{i=1,j=1,\ i=k,j<>k}^{n} a_{ij}\,x_ix_j = \sum_{j=1,\ j<>k}^{n} a_{kj}\,x_kx_j = \sum_{j=1,\ j<>k}^{n} a_{kj}\,x_jx_k$

3. $\sum_{i=1,j=1,\ i<>k,j=k}^{n} a_{ij}\,x_ix_j = \sum_{i=1\ mit\ i!=k}^{n} a_{ik}\,x_kx_j = \sum_{i=1,\ i<>k}^{n} a_{ik}\,x_jx_k$

4. $\sum_{i=k,j=k}^{n} a_{ij}\,x_ix_j = a_{kk}x_k^2$

We will decompose the term $a_{ij}x_ix_j$ as a sum and get:

$$a_{ij}x_ix_j \equiv \sum_{i=1,j=1,\ i<>k,j<>k}^{n} a_{ij}x_ix_j \;+\; \sum_{j=1,\ j<>k}^{n} a_{kj}x_jx_k$$

$$+\; \sum_{i=1,\ i<>k}^{n} a_{ik}x_jx_k \;+\; \sum_{i=k,j=k}^{n} a_{ij}x_ix_j$$

With the preliminary remarks we get:

$$a_{ij}x_ix_j \equiv K \;+\; \sum_{j=1,\ j<>k}^{n} a_{kj}x_jx_k \;+\; \sum_{i=1,\ i<>k}^{n} a_{ik}x_jx_k \;+\; a_{kk}x_k^2$$

where K is a constant independent from x_k.

Now we take the partial derivative:

$$\frac{\partial}{\partial x_k}\left(a_{ij}x_ix_j\right)$$

$$\equiv \frac{\partial}{\partial x_k}\left(K + \sum_{j=1,j!=k}^{n} a_{kj}x_jx_k + \sum_{i=1,\,i!=k}^{n} a_{ik}x_jx_k + a_{kk}x_k^2\right)$$

$$= 0 + \sum_{j=1,\,j!=k}^{n} a_{kj}x_j + \sum_{i=1,\,i!=k}^{n} a_{ik}x_i + 2a_{kk}x_k$$

$$= \sum_{j=1}^{n} a_{kj}x_j + \sum_{i=1}^{n} a_{ik}x_i \qquad\qquad (*)$$

$$\equiv a_{ki}x_i + a_{ik}x_i$$

$$= (a_{ik} + a_{ki})x_i$$

(*) *the conditions j !=k and i != k do not apply any more because we added $a_{kk}\,x_k$ in each sum.*

The equation can also be viewed more easily by applying the product rule to derivatives and observing (PA.0):

$$\frac{\partial}{\partial x_k}\left(a_{ij}x_ix_j\right) = a_{ij}\,\frac{\partial}{\partial x_k}\left(x_ix_j\right)$$

$$= a_{ij}\left(x_j\,\frac{\partial}{\partial x_k}(x_i) + x_i\,\frac{\partial}{\partial x_k}(x_j)\right)$$

$$= a_{ij}\left(x_j\delta_{ik} + x_i\delta_{jk}\right)$$

$$= a_{ij}x_j\delta_{ik} + a_{ij}x_i\delta_{jk}$$

$$= a_{ij}\delta_{ik}x_j + a_{ij}\delta_{jk}x_i$$

$$= a_{kj}x_j + a_{ik}x_i$$

$$= (a_{ik} + a_{ki})x_i$$

(PA.4) $\qquad \dfrac{\partial}{\partial x_l}\left(a_{ijk}x_i x_j x_k\right) = \left(a_{lij} + a_{ilj} + a_{ijl}\right)x_i x_j$

We have:

$$
\begin{aligned}
\frac{\partial}{\partial x_l}\left(a_{ijk}x_i x_j x_k\right) \;=\;& a_{ijk}\frac{\partial}{\partial x_l}\left(x_i x_j x_k\right)\\[2mm]
=\;& a_{ijk}\left[\,x_i\frac{\partial}{\partial x_l}\left(x_j x_k\right) + x_j x_k\frac{\partial}{\partial x_l}\left(x_i\right)\,\right]\\[2mm]
=\;& a_{ijk}\left[\,x_i\left(x_j\frac{\partial}{\partial x_l}\left(x_k\right) + x_k\frac{\partial}{\partial x}\left(x_j\right)\right)\right.\\
& \left.\qquad\qquad + x_j x_k\delta_{il}\,\right]\\[2mm]
=\;& a_{ijk}\left[x_i x_j\delta_{lk} + x i x k\delta_{lj} + x_j x_k\delta_{il}\right]\\[2mm]
=\;& a_{ijl}x_i x_j + a_{ilk}x_i x_k + a_{ljk}x_j x_k
\end{aligned}
$$

We consider the first term $a_{ijl}x_i x_j$: here we can exchange a_{ijl} by a_{lij} .

In the term $a_{ilk}x_i x_k$ we replace the index "k" by "j". We may do this as "k" (as well as "i") are summation indices (cf. (E.0)).

Finally, we exchange in the last term $a_{ljk}x_j x_k$ the indexes as follows: "k" becomes "j" and "j" becomes "i".

Putting this together, we get the assertion.

ESN AND DIFFERENTIAL OPERATORS

In the following we consider differential operators. To do this, we first give the definition of the differential operator and then see how this is represented in the ESN formulation. After that, we will prove some basic connections between the differential operations using the ESN notation. In these proofs we will use ε_{ijk} and see that the individual proofs can be carried out elegantly, briefly and clearly.

In the following we always move in Euclidean space $\mathbb{R}^n$ with the unit base $\left(\vec{e_1}, \vec{e_2}, \ldots, \vec{e_n}\right)$. We justify this by saying that with this definition of space, the differential operators to be considered assume a simple form. Gradient, divergence and curl can also be represented with the help of certain integrals independently of the selected coordinate system, but in such a representation it is not possible to work with the ESN or only to a very limited extent.

We consider two differentiable functions (they should be at least 2-fold differentiable):

$$S:\mathbb{R}^n \to \mathbb{R} \qquad \text{(scalar field, a vector is assigned a real number) as well as}$$

$$\vec{V}:\mathbb{R}^n \to \mathbb{R}^n \qquad \text{(Vector field, a vector is assigned to a vector)}$$

$\vec{V}$ can also be considered componentwise as a vector field: $\vec{V} = (V_1, V_2, \ldots, V_n)$, we will need this below.

It is important to understand the difference between a scalar field and a vector field: a scalar field maps a vector to a real number, while a vector field maps a vector to a vector.

Nabla-Operator ∇

The Nabla-Operator ∇ is a vector whose components are the partial derivatives:

$$\nabla = \begin{pmatrix} \dfrac{\partial}{\partial x_1} \\[2ex] \dfrac{\partial}{\partial x_2} \\ \cdots \\ \dfrac{\partial}{\partial x_n} \end{pmatrix}$$

The scalar product of ∇ with itself is defined as the **Laplace-operator**:

$$\Delta = \nabla^2 = \nabla \cdot \nabla$$

We can use the Nabla-operator to form various types of products, which we will consider below.

Gradient of a scalar field

The gradient of a scalar field S is defined as follows:

$$grad\ S = \begin{pmatrix} \dfrac{\partial S}{\partial x_1} \\ \dfrac{\partial S}{\partial x_2} \\ \ldots \\ \dfrac{\partial S}{\partial x_n} \end{pmatrix}$$

The gradient of a scalar field S is therefore a vector whose components consist of the partial derivatives of S.

As already indicated, the gradient S can be written as a scalar product using the Nabla operator:

$$grad\ S = \nabla \cdot S$$

In ESN notation we get:

(GRAD.1) $grad\ S = \vec{e_i}\,\dfrac{\partial S}{\partial x_i}$

It should be noted that the index i appears twice and is therefore summed over it. The gradient is NOT the sum of the partial derivatives - realize that the notation of (GRAD.1) refers to components to the basis $\left(\vec{e_1}, \vec{e_2}, \ldots, \vec{e_n}\right)$.

Divergence of a vector field

The scalar product of ∇ with a vector field gives it's divergence:

$$\nabla \cdot \vec{V} = div\ \vec{V} = \sum \frac{\partial \vec{V}}{\partial x_i}$$

In ESN notation:

(DIV.1) $\nabla \cdot \vec{V} = div\ \vec{V} = \dfrac{\partial \vec{V_i}}{\partial x_i}$

In contrast to the gradient, here we have an actual summation of the partial derivatives of the individual component functions of the vector field.

Curl

The cross product of with a vector field V defines the curl:

$$curl\ \vec{V} = \nabla \times \vec{V} = \begin{pmatrix} \dfrac{\partial V_3}{\partial x_2} - \dfrac{\partial V_2}{\partial x_3} \\[2mm] \dfrac{\partial V_1}{\partial x_3} - \dfrac{\partial V_3}{\partial x_1} \\[2mm] \dfrac{\partial V_2}{\partial x_1} - \dfrac{\partial V_1}{\partial x_2} \end{pmatrix}$$

This is only valid in the 3-dimensional Euclidean space.

In ESN we get "the easy to check" notation:

(CURL.1) $curl\ \vec{V} = \vec{e_i}\varepsilon_{ijk} \dfrac{\partial V_k}{\partial x_j}$

Note that we sum over i, j and k here.

We will now derive some known equations and relationships between gradients, divergence and curl using index-notation:

(DIFF.1) $curl\ grad\ S = 0$

We have:

$$
\begin{aligned}
curl\ grad\ S\ =\ & curl\ \vec{e}_k\,\frac{\partial S}{\partial x_k} \\[2mm]
=\ & curl\ \vec{e}_1\,\frac{\partial S}{\partial x_1} + curl\ \vec{e}_2\,\frac{\partial S}{\partial x_2} + curl\ \vec{e}_3\,\frac{\partial S}{\partial x_3} \\[2mm]
=\ & \vec{e}_1\varepsilon_{1jk}\,\frac{\partial^2 S}{\partial xj\,\partial x_k} + \vec{e}_2\varepsilon_{2jk}\,\frac{\partial^2 S}{\partial xj\,\partial x_k} + \vec{e}_3\varepsilon_{3jk}\,\frac{\partial^2 S}{\partial xj\,\partial x_k} \\[2mm]
\overset{(1)}{=}\ & \vec{e}_1\left(\varepsilon_{123}\,\frac{\partial^2 S}{\partial x_2\,\partial x_3} + \varepsilon_{132}\,\frac{\partial^2 S}{\partial x_3\,\partial x_2}\right) \\[2mm]
+\ & \vec{e}_2\left(\varepsilon_{231}\,\frac{\partial^2 S}{\partial x_3\,\partial x_1} + \varepsilon_{213}\,\frac{\partial^2 S}{\partial x_1\,\partial x_3}\right) \\[2mm]
+\ & \vec{e}_3\left(\varepsilon_{312}\,\frac{\partial^2 S}{\partial x_1\,\partial x_2} + \varepsilon_{321}\,\frac{\partial^2 S}{\partial x_2\,\partial x_1}\right) \\[4mm]
=\ & \vec{e}_1\left(1\,\frac{\partial^2 S}{\partial x_2\,\partial x_3} + (-1)\,\frac{\partial^2 S}{\partial x_3\,\partial x_2}\right) \\[2mm]
+\ & \vec{e}_2\left(1\,\frac{\partial^2 S}{\partial x_3\,\partial x_1} + (-1)\,\frac{\partial^2 S}{\partial x_1\,\partial x_3}\right) \\[2mm]
+\ & \vec{e}_3\left(1\,\frac{\partial^2 S}{\partial x_1\,\partial x_2} + (-1)\,\frac{\partial^2 S}{\partial x_2\,\partial x_1}\right) \\[4mm]
=\ & 0
\end{aligned}
$$

On the one hand, we took advantage of that ε_{ijk} becomes 0 with two equal indices (step (1)). In the penultimate step, we used Schwartz's theorem, according to which the order of multiple partial derivatives can be exchanged as desired.

(DIFF.2) $div\ \ curl\ \vec{V} = 0$

We have:

$$div\ \ curl\ \vec{V}\ =\ div\ \vec{e_i}\ \epsilon_{ijk}\ \frac{\partial V_k}{\partial x_j} \qquad [\,(CURL.1)\,]$$

$$=\ \frac{\partial V_i}{\partial x_i}\ \epsilon_{ijk}\ \frac{\partial V_k}{\partial x_j} \qquad [\,(DIV.1)\,]$$

$$=\ \epsilon_{ijk}\ \frac{\partial V_i}{\partial x_i}\ \frac{\partial V_k}{\partial x_j} \qquad [\,reorder\,]$$

$$=\ 0 \qquad [\,Form\ summation\ as\ in\ proof\ (DIFF.1)\,]$$

The fact that the last sum is equal to zero can be seen with the same argument as was already explained in the previous proof: in the case of the direct calculation of the sum, one only has to worry about those cases where there are no two identical indices in ε_{ijk} are used. What is then left over disappears with Schwartz's theorem.

(DIFF.3) $curl\ curl\ \vec{V} = grad\ div\ \vec{V} - \Delta\ \vec{V}$

We first note that by definition (CURL.1) the k-th component of $rot\ \vec{V}$ can be written as:

$$\left[curl\ \vec{V} \right]_k = \vec{e}_k\ \epsilon_{klm}\ \frac{\partial V_m}{\partial x_l}$$

Keeping this in mind, we get:

$$
\begin{aligned}
curl\ curl\ \vec{V}\ &=\ \vec{e}_i\ \epsilon_{ijk}\ \frac{\partial}{\partial x_j} \left[rot\ \vec{V} \right]_k && def \\[2ex]
&=\ \vec{e}_i\ \epsilon_{ijk}\ \frac{\partial}{\partial x_j}\ \epsilon_{klm}\ \frac{\partial}{\partial x_l} V_m && def,\ (\text{E.7}) \\[2ex]
&=\ \vec{e}_i\ \epsilon_{kij}\ \frac{\partial}{\partial x_j}\ \epsilon_{klm}\ \frac{\partial}{\partial x_l} V_m && (\text{LC.1}) \\[2ex]
&=\ \vec{e}_i\ \epsilon_{kij}\ \epsilon_{klm}\ \frac{\partial}{\partial x_j}\frac{\partial}{\partial x_l} V_m && \\[2ex]
&=\ \vec{e}_i \left(\delta_{il}\,\delta_{jm} - \delta_{im}\,\delta_{jl} \right) \frac{\partial}{\partial x_j}\frac{\partial}{\partial x_l} V_m && (\text{LC.6}) \\[2ex]
&=\ \vec{e}_i \left(\delta_{il}\,\delta_{jm}\ \frac{\partial}{\partial x_j}\frac{\partial}{\partial x_l} V_m - \delta_{im}\,\delta_{jl}\ \frac{\partial}{\partial x_j}\frac{\partial}{\partial x_l} V_m \right) && contract \\[2ex]
&=\ \vec{e}_i \left(\frac{\partial}{\partial x_m}\frac{\partial}{\partial x_i} V_m - \frac{\partial}{\partial x_l}\frac{\partial}{\partial x_l} V_i \right) && \\[2ex]
&=\ \vec{e}_i\ \frac{\partial}{\partial x_m}\frac{\partial}{\partial x_i} V_m - \vec{e}_i\ \frac{\partial}{\partial x_l}\frac{\partial}{\partial x_l} V_i && \\[2ex]
&=\ \vec{e}_i\ \frac{\partial}{\partial x_i} div\ V\ - \nabla^2 \vec{V} && \\[2ex]
&=\ grad\ div\ \vec{V} - \Delta\ \vec{V} &&
\end{aligned}
$$

(DIFF.4) Let S_1 and S_2 be two scalar fields, then the following equation holds:

$$grad \ S_1 S_2 = S_1 \ grad \ S_2 + S_2 \ grad \ S_1$$

We have:

$$
\begin{aligned}
grad \ S_1 S_2 \ &= \ \vec{e_i}\left(\frac{\partial \ S_1 S_2}{\partial x_i}\right) && [\ Def.\ of\ grad\ ] \\[2mm]
&= \ \vec{e_i}\left(S_1 \frac{\partial \ S_2}{\partial x_i} + S_2 \frac{\partial \ S_1}{\partial x_i}\right) && [\ chain\ rule\ ] \\[2mm]
&= \ \vec{e_i}\, S_1 \frac{\partial \ S_2}{\partial x_i} + \vec{e_i}\, S_2 \frac{\partial \ S_1}{\partial x_i} && [\ multiply\ out\ ] \\[2mm]
&= \ S_1 \ grad \ S_2 + S_2 \ grad \ S_1 && [\ apply\ def.\ ]
\end{aligned}
$$

(DIFF.5) Let **V1** and **V2** be two vector fields, then:

$$curl \ (V_1 \times V_2) = (V_2 \ grad) \ V_1 - (V_1 \ grad \ )V_2 + V_1 \ div \ V_2 - V_2 \ div \ V_1$$

We have:

$$
curl \ \left(\vec{V_1} \times \vec{V_2}\right) \ = \ \vec{e_i}\, \epsilon_{ijk}\, \frac{\partial}{\partial x_j}[\, V_1 \times V_2]_k \tag{1}
$$

$$
= \ \vec{e_i}\, \epsilon_{ijk}\, \frac{\partial}{\partial x_j}[\, V_1 \times V_2]_k \tag{2}
$$

$$
= \ \vec{e_i}\, \epsilon_{ijk}\, \frac{\partial}{\partial x_j}[\, V_1 \times V_2]_k \tag{3}
$$

$$= \quad \vec{e_i} \, \frac{\partial}{\partial x_j} \, (\delta_{il}\delta_{jm} - \delta_{im}\delta_{jl})[V_1]_l[V_2]_m \tag{4}$$

$$= \quad \vec{e_i} \, \frac{\partial}{\partial x_j} \, (\delta_{il}\delta_{jm} \, [V_1]_l[V_2]_m - \delta_{im}\delta_{jl} \, [V_1]_l[V_2]_m) \tag{5}$$

$$= \quad \vec{e_i} \, \frac{\partial}{\partial x_j} \, ([V_1]_i[V_2]_j) - \vec{e_i} \, \frac{\partial}{\partial x_j} \, ([V_1]_j[V_2]_i) \tag{6}$$

$$\begin{aligned}
= \quad & \vec{e_i} \, [V_2]_j \, \frac{\partial}{\partial x_j} \, ([V_1]_i) + \vec{e_i} \, [V_1]_i \, \frac{\partial}{\partial x_j} \, ([V_2]_j) \\
& - \vec{e_i} \, [V_1]_j \, \frac{\partial}{\partial x_j} \, ([V_2]_i) \\
& - \vec{e_i} \, [V_2]_i \, \frac{\partial}{\partial x_j} \, ([V_1]_j)
\end{aligned} \tag{7}$$

$$\begin{aligned}
= \quad & [V_2]_j \, \frac{\partial}{\partial x_j}\left([V_1]_i \, \vec{e_i}\right) + \vec{V_1} \, div \, \vec{V_2} \\
& - [V_1]_j \, \frac{\partial}{\partial x_j}\left([V_2]_i \, \vec{e_i}\right) \\
& - \vec{V_2} \, div \, \vec{V_1}
\end{aligned} \tag{8}$$

$$\begin{aligned}
= \quad & \left(\vec{V_2} \, grad \, \right) \vec{V_1} + \vec{V_1} \, div \, \vec{V_2} - \left(\vec{V_1} \, grad \, \right) \vec{V_2} \\
& - \vec{V_2} \, div \, \vec{V_1}
\end{aligned} \tag{9}$$

(1)   Def. of curl (2) Def. of cross-product, (E.7) (3) Reorder (4) apply (LC.6) (5) multiply out

(6)   Contraction (7) chain rule (8) reorder (9) apply definition

Summary

In the following, the symbolic notation is compared with the ESN notation and component notation in the ESN. Let $\vec{v}$ and $\vec{w}$ be vectors with the components $(v_1, v_2, \ldots, v_n)$ resp. $(w_1, w_2, \ldots, w_n)$ and $\left(\vec{g}_1, \vec{g}_2, \ldots, \vec{g}_n\right)$ be a basis.

Symbolic Notation	ESN-Notation	ESn-Notation in components
v	$v_i\, \vec{g}_i$	$[v]_i = v_i$
$v \cdot w$	$v_i w_i$	$v_i w_i$
$v \otimes w$	$v_i\, w_j\, \vec{g}_i \otimes \vec{g}_j$	$[v \otimes w]_{ij} = v_i w_j$
$v \times w$	$v_i\, w_j\, \epsilon_{ijk}\, \vec{g}_k$	$[v \times w]_k = v_i w_j \varepsilon_{ijk}$
$v \cdot (w \times y)$	$v_i w_j y_k \varepsilon_{ijk}$	$v_i w_j y_k \varepsilon_{ijk}$

BIBLIOGRAPHY

There is surprisingly little literature on the ESN - mostly it is introduced in just a few sentences.

Schaum's Outlines: David C.Kay: Tensor Calculus

Provides a good introduction to ESN with numerous examples. Anyone who wants to deal with tensors is well advised here.

Björn Feuerbacher: Tutorium Mathematische Methoden der Elektrodynamik

(In german!) Although ESN is explicitly not used in this book, numerous divergence, curl, and gradient equations are presented in "almost" ESN notation.

There is a wealth of information about the ESN on the internet. From the many sources, the following are listed:

Alan H. Barr: The Einstein Summation Notation

http://vucoe.drbriansullivan.com/wp-content/uploads/Einstein-Summation-Notation.pdf

The script describes some interesting suggestions for extensions to the ESN. These extensions also allow for index notations that could not previously be represented as ESN. As an example, the extensions are considered using quaternions.

Albert Einstein: Die Grundlage der allgemeinen Relativitätstheorie

http://www.alberteinstein.info/gallery/pdf/CP6Doc30_pp284-339.pdf

(In german) In this original edition one can see how A. Einstein uses the summation named after him.